Counting Around Town

COUNTING AT THE AQUARIUM

Rosie Banks

PowerKiDS press

PK Beginners

We're at the aquarium.
I count animals
in the water!

I count one sea otter.
It has thick fur.

1

I count two
whale sharks.
They are so big!

1
2

I count three penguins.
They cannot fly.

1
2
3

I count four seahorses. They live in warm water.

1
2
3
4

I count five sea lions.
They like to play!

1
2
3
4
5

I count six dolphins.
They like to jump!

1
2
3
4
5
6

I count seven jellyfish.
They have no bones!

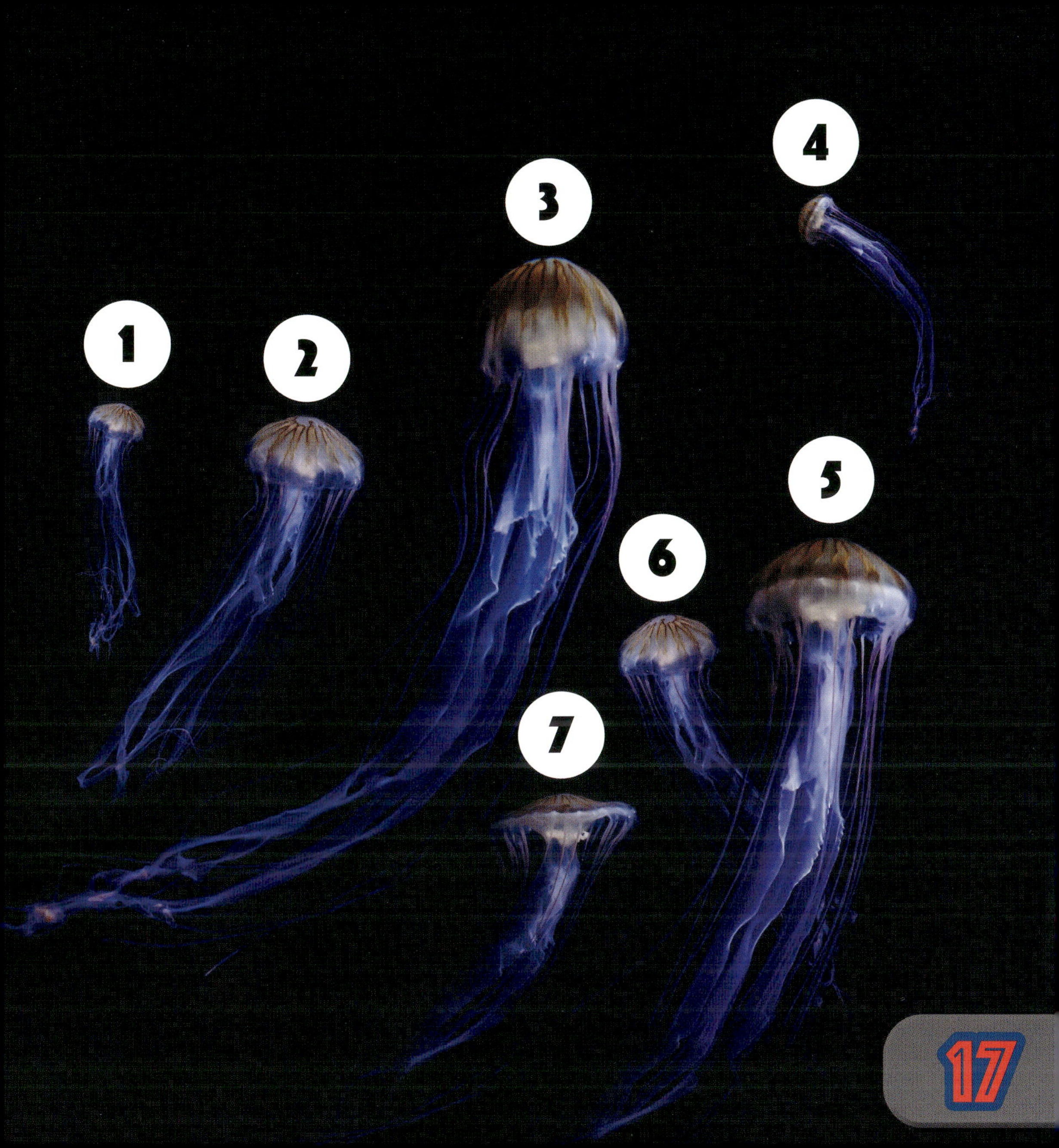
1
2
3
4
5
6
7

I count eight legs.
It’s an octopus!

1
2
8
3
7
4
6
5

I count nine sea stars. They live on the ocean floor.

1
2
3
4
5
6
7
8
9

How many clown fish
do you count?
I count ten!

1
2
3
4
5
6
7
8
9
10

Published in 2025 by The Rosen Publishing Group, Inc.
2544 Clinton Street, Buffalo, NY 14224

First Edition

Editor: Therese M. Shea
Book Design: Rachel Rising

Photo Credits: Cover, p.1 lingling7788/Shutterstock.com; p. 3 JaySi/Shutterstock.com; p. 5 Erin Underwood/Shutterstock.com; p. 7 f11photo/Shutterstock.com; p. 9 Grzegorz Kandzia/Shutterstock.com; p. 11 Studio 37/Shutterstock.com; p. 13 VIAVAL TOURS/Shutterstock.com; p. 15 Grommik/Shutterstock.com; p. 17 chonlasub woravichan/Shutterstock.com; p. 19 Vladimir Wrangel/Shutterstock.com; p. 21 Damsea/Shutterstock.com; p. 23 Independent birds/Shutterstock.com.

Library of Congress Cataloging-in-Publication Data

Names: Banks, Rosie, 1978- author.
Title: Counting at the aquarium / Rosie Banks.
Description: New York : PowerKids Press, (2025) | Series: Counting around town
Identifiers: LCCN 2023051703 (print) | LCCN 2023051704 (ebook) | ISBN 9781499444759 (library binding) | ISBN 9781499444742 (paperback) | ISBN 9781499444766 (ebook)
Subjects: LCSH: Counting--Juvenile literature. | Aquariums--Juvenile literature. | Aquatic animals--Juvenile literature.
Classification: LCC QA113 .B354 2025 (print) | LCC QA113 (ebook) | DDC 513.2/11--dc23/eng/20231124
LC record available at https://lccn.loc.gov/2023051703
LC ebook record available at https://lccn.loc.gov/2023051704

Manufactured in the United States of America

CPSIA Compliance Information: Batch #CSPK25. For further information contact Rosen Publishing at 1-800-237-9932.